AF604552

DEADLY SCIENCE

Light and colour

Contents

ADJUNCT ASSOCIATE PROFESSOR COREY TUTT OAM

DEADLYSCIENCE

DeadlyScience aims to provide Science, Technology, Engineering and Mathematics (STEM) resources to remote schools around Australia. So far, DeadlyScience has shipped more than 33,000 STEM books and resources to more than 800 schools across the country.

The organisation began when proud Kamilaroi man Corey Tutt found out that some schools in Australia were completely under-resourced and that Aboriginal and Torres Strait Islander children were discouraged from pursuing STEM because of this. DeadlyScience knows from personal experience that books and resources change lives and believes these kids deserve nothing but the best. Aboriginal and Torres Strait Islander peoples in Australia were the First Scientists of this land, and DeadlyScience is committed to preserving that history.

Lighting our way

Light has always played a huge role in human lives – and it always will because we could not survive without the Sun's light and warmth. Harnessing light and heat allowed people to stay warm, cook food and see more clearly at night by creating fires and torches. From ancient campfires to lamps and modern lasers, over millennia humans have searched for ways to create light, and, with it, energy. Light also produces the many wonderful hues of the world around us, from the colour of your skin and eyes to a beautiful sunset, a rainbow or the twinkling stars in the sky.

MAKING FIRE

First light

Today, people use light in amazing devices, including sensors to measure time, microscopes, telescopes, and even lasers to guide rocket propulsion! So exactly how does light work? To figure that out, scientists first had to peer all the way back to the dawn of time itself.

Born with a bang

The Big Bang is a scientific theory to explain the origins of our universe. Some 13.8 billion years ago, a massive explosion of energy and matter rapidly spread out into our universe.

The early stages of expansion generated a tremendous amount of heat and light energy, encouraging charged particles to group together. As that cycle continued, particles joined to form the galaxies, stars, suns and solar systems we know today. When you look at a star, you're really looking at a giant ball of fire (just like the star that is our Sun). What you don't see are millions upon millions of tiny particles fused together by the star's energy, releasing even more energy in the form of heat and light.

FUSION GIVES OFF LIGHT AND HEAT

FACT

Fusion is the process of multiple particles becoming one, and it creates light from sunrise to sunset.

The first glow sticks

Fire is a natural process and First Australians have been using it for light, heat, cooking and managing Country for as far back as their history is known.

To make fire, First Australians traditionally collect thin wood shavings from bark or other fibres that burn easily. Taking two dry sticks, they place one down on the ground and create a shallow groove in it. Then, they put the flammable material and the end of the second stick, held vertically, into the groove and rapidly roll the vertical stick between both hands, making it turn back and forth. These quick movements make kinetic energy – energy due to motion – which transforms into heat. The faster the stick spins, the more heat it makes, until the fine wooden gauze starts to smoke and form embers. Blowing gently on the embers encourages a fire to start.

FACT

The wedge-tailed eagle is a totem animal for the Awabakal people of Newcastle and Lake Macquarie, NSW.

DID YOU KNOW?

First Australians have many explanations for the creation of our world. One, from the Wurundjeri people of the Yarra River Valley, Victoria, tells how the world came to be when a wedge-tailed eagle creator being known as Bunjil created mountains, rivers, humans and animals before asking Bellin-bellin, the musk crow, who was in charge of winds, to blow him and his family up into the sky where they now look down on the world as stars.

WEDGE-TAILED EAGLE

This raptor is known as birabaan to the Awabakal people and bunjil to the Wurundjeri people.

Building blocks of life

Fusion allows microscopic particles to form larger particles known as atoms, and, just like building something out of blocks or Lego® bricks, smaller atoms fuse into bigger atoms, and so on, which create the planets, moons and other masses floating through space. For instance, two different forms, or isotopes, of hydrogen fuse together to become helium gas, releasing a large amount of energy in the process.

How fusion works

1. Hydrogen atoms are heated.
2. Fusion reaction takes place when two isotopes of hydrogen (H) known as deuterium and tritium fuse together to make helium (He).
3. Helium gas, neutrons and energy are all released.
4. Neutron energy can be used to heat water and produce electricity.

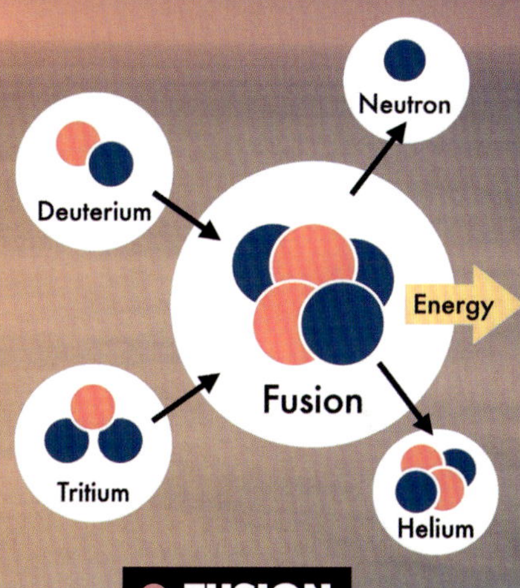

FUSION

BIG BANG EXPANSION

How it happened...

1. In the first three minutes after the Big Bang, our universe began as atomic particles.
2. Cooling caused clouds of gas.
3. Lumps of gas became the first stars.
4. Gas clouds collided to form galaxies.
5. Spinning discs of gas and dust formed stars and planets.

Nature's light show

Light is a form of energy like heat or sound. But not all light comes from the Sun. Nature offers other amazing and beautiful examples of light all around us. Other forms of energy include electrical energy, nuclear energy, kinetic energy, gravitational potential energy, and the list goes on. In physics, energy is defined as a measure of an object's ability to do work; that is, if it can make something happen, it has energy. Chemical energy is also created from the bonds between particles, and it sets off chain reactions within plants and animals, too, resulting in bioluminescence.

FACT

Auroras (the southern and northern lights) result from charged particles from the Sun's solar wind interacting with gases in Earth's atmosphere and magnetic field.

SOUTHERN LIGHTS

All the light we cannot see

We don't just get visible light from our Sun, we also get charged particles in the form of solar wind, heat, magnetic and gravitational fields, and a range of energy we can't see, including ultraviolet (UV) radiation. Even the light we can see has travelled close to 150,000,000 km to our eyes. Approximately 48% of the Sun's energy makes it through our atmosphere to reach us on Earth, the rest either bounces off or gets absorbed. Imagine how fiery the Sun must look from space!

Moonlight, moon bright

Depending on the position of the Earth, Sun and Moon, we see a different part of the Moon each night. What we're really seeing is sunlight bouncing off the Moon's stark surface and reflecting the Sun's light back to us. The image below demonstrates how the light casts a shadow on the Moon, which creates the Moon's waxing and waning phases we observe from the Earth. The side of the Moon we don't see is known as the lunar 'far side', and the first images of it were taken in 1959.

DID YOU KNOW?

The Moon reflects the Sun's light, which is 100,000 times brighter. But the light beams produced by particle accelerators at CERN are over 1,000,000 times brighter than the Sun!

MOON PHASES

BIOLUMINESCENT GHOST FUNGI

DID YOU KNOW?

You may see bioluminescent trails left behind in the sand at night by miniscule crustaceans called ostracods. The Yolngu people believe these trails represent the the Djang-kawu sisters from the creation stories of the Dhuwa group, which link the natural world to the Dreaming, reinforcing the strong connection between the people and their ancestors.

AN OSTRACOD
seen under a polarised microscope.

A biological glow

Bioluminescence occurs when a living organism creates its own light through a chemical reaction within its cells. This reaction between certain enzymes (such as luciferase) and oxygen excites molecules, which release excess energy in the form of visible light.

Deep-sea fish, fireflies, glow worms and even the 'sea sparkles' caused by bioluminescent *Noctiluca scintillans* plankton are examples of bioluminescence. Typically, these organisms use their light to defend against predators, entice prey or attract mates. Sometimes, bioluminescence is the product of a symbiotic relationship between two organisms, such as the tiny glowing organisms that light up the 'lure' of an anglerfish.

BIOFLUORESCENT CORAL

Biofluorescence

If you've ever snorkelled, no doubt you'll have seen extremely colourful fish and coral. When exposed to types of light, tiny micro-organisms inside them absorb and re-emit that glow, often creating vivid displays. These life forms rarely create their own light – they just reuse it. Absorbing light of one wavelength (often ultraviolet or dark blue light) and re-emitting it as another is called biofluorescence. For example, chlorophyll, a vital chemical that helps plants convert sunlight into food, is used by some deep sea algae to emit a bright-red colour. Scientists can determine the health of many types of plants based on the colour they emit under certain light conditions in a lab.

SEA SPARKLES

FACT

The fur of some of Australia's unique mammal species, including the Tasmanian devil and the platypus, biofluoresces under ultraviolet light.

Light & energy

FACT

You're a star! All the particles that make up our planet were once parts of a star, which means you are made up of stardust, too.

ÉMILIE DU CHÂTELET

Light, like all energy, must come from somewhere. In the 18th century, French philosopher and mathematician Émilie du Châtelet formed the basis of the law of the conservation of energy, which was later expanded upon by Isaac Newton and Albert Einstein. Basically, it recognises that energy cannot be created from nothing, nor can it disappear. Energy can only be transferred (passed along to another object) or transformed (converted into a different type of energy).

An excess of energy

Light is created when there is too much energy and some needs to be released. Natural light is most often excess chemical energy being released as light energy; however, there are many other ways that transformation of energy can take place.

DID YOU KNOW?

In the Noongar language of WA, ngangk means both Sun and mother.

CHEMICAL ENERGY

KINETIC ENERGY

GRAVITATIONAL POTENTIAL ENERGY

SOUND & HEAT ENERGY

Energy transformation

When you pick up a ball and bounce it or hit it with a tennis raquet, much more is going on than you think. Let's consider the chain reaction of energy involved...

1 We consume chemical energy in the food we eat.

2 That chemical energy is transformed into kinetic energy (our motion) as we move to pick up the ball. In turn, energy is transferred into the ball in the form of gravitational potential energy (the ability for gravity to work is proportional to how high a thing is).

3 Then, once the ball falls, it converts that energy back into kinetic energy as it speeds up.

4 Eventually the ball hits the ground, producing energy in the form of sound and sometimes heat.

FACT

Whenever an object's speed doubles, its kinetic energy increases by four!

Artificial light

Natural light is usually chemical energy being converted and released as light energy, but humans learned how to create artificial light by forcing energy to transform. We found out how to make lasting light sources, such as torches or candles, and we discovered that natural oils in wax, melted-down animal fats like whale oil, and coal oil or kerosene could feed a flame at the end of a wick.

Let there be light!

Humans figured out how to generate electrical energy when Joseph Swan and Thomas Edison invented the incandescent light bulb in 1878. It produced light from a current of electricity running through a thin piece of wire known as a filament. The wire heated up and released excess energy in the form of light. However, the amount of energy lost to heat makes incandescent bulbs both inefficient and hot to the touch.

DID YOU KNOW?

The Mer people of the Torres Strait Islands used fronds from coconut palms as torches at night.

FLUORESCENT BULB

INCANDESCENT BULB

Going fluorescent

Fluorescent lamps later replaced incandescent bulbs, and they work in two stages. The first uses a current to give energy to a mercury vapour inside a tube, resulting in highly energised mercury atoms. The electrons in the vapour then lose their energy again, releasing it as packets of light called photons. As the energy of those molecules is released in the form of visible and UV light, it interacts with another chemical, phosphor, which coats the tube. So the second stage is that phosphor converts the light into the bright visible light of fluorescent lamps.

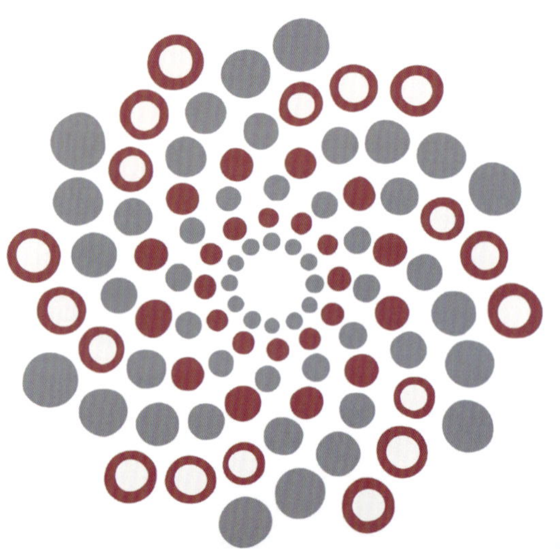

Neon lights

Neon lights are usually long tubes of neon gas bent into a desired shape, such as a word or a picture. When electricity passes through the neon gas, it excites the atoms, which release energy in the form of very bright light in vivid colours. You often see these lights in shop windows or signs that are designed to draw our attention.

LED LIGHTS

LED lights

LED is an acronym for light-emitting diode. Diodes are devices that only allow electricity to flow in one direction, a bit like a one-way street.

Similar to light bulbs, the current excites electrons, which then release energy in the form of light. What's special about LEDs is that, because they are components in electrical circuits, they can be very small and produce little heat, making them useful for everything from pen lights to the headlights on your car. Modern applications of LEDs use UV light to detect counterfeit money and even to sterilise water!

Mysterious Min Mins

In parts of outback Australia, strange lights are sometimes seen in the distance, appearing to dance along the ground or follow travellers at night. These Min Min lights may be a trick of the light caused by refraction (see page 18), or may even be a figment of the imagination, but they feature in Aboriginal storytelling across many groups, and Indigenous children are often taught about them.

Some believe the lights to be Elders past, roaming the land and looking after Country. Others believe they are angry spirits who were banished and are trying to find a way back.

The presence of Min Min lights in Indigenous lore, along with their regular sightings, suggests the possibility of an interesting discovery for the next generation of young scientists.

Light waves

To understand how light interacts with the world, we first need to understand waves. Light waves work in the same way that ocean waves or sound waves do. The wave model for light was first accepted in 1690 after being submitted by a Dutch mathematician by the name of Christiaan Huygens. Over the years, his model for explaining light has evolved, but it is still the most widely accepted explanation for the behaviour of light.

What makes waves?

Every wave has three key components.

1 WAVELENGTH

If you think of waves at the beach, wavelength refers to the distance between the peaks or troughs of each wave. In physics, wavelength is represented by the Greek letter lambda (λ) and is measured in metres. In light waves, the electromagnetic spectrum encompasses a range of wavelengths, from long radio waves to short gamma rays. The visible light our eyes can see spans a specific range of wavelengths from approximately 380 to 700 nanometres. A single nanometre is equal to 0.000000001 metres, or one metre divided in a billion pieces! If you know the colours of the rainbow (red, orange, yellow, green, blue, indigo, and violet), you also know each colour in the order of their wavelength, from longest to shortest.

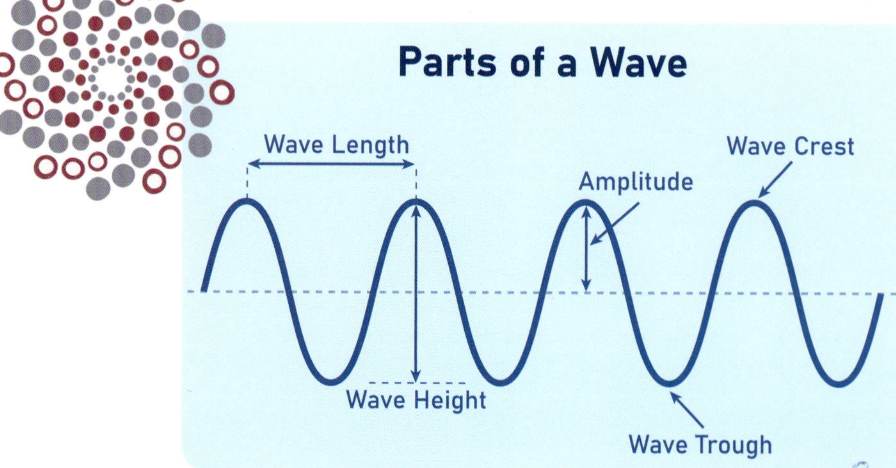

2 FREQUENCY

Frequency refers to how often we observe a full wave (peak and trough) passing a single point in one second. It depends on how fast the waves are moving and how long they are (their wavelength). If you imagine two trucks, bumper to bumper, passing at 60 km/h, it might take them one second to go past – a frequency of two. But if there were smaller cars all lined up and moving at the same speed, as many as six cars might go past in the same time. The shorter cars would create a higher frequency. An equation links the relationship between frequency, wavelength and velocity (speed).

$$\mathbf{v} = f \times \lambda$$

velocity (metres per second) = frequency (waves per second) × wavelength (metres)

3 AMPLITUDE

The amplitude of a wave is simply how high or low it gets in comparison to its most central position. If you imagine a cork in the ocean, floating as each crest and trough rolls past out beyond the breakers, the amplitude refers to how much higher or how much lower the cork gets compared to when the water is still and flat.

When talking about light, the higher the amplitude, the brighter the light. This is the same for sound. Increasing the volume produces the same musical note (frequency), but much louder (amplitude).

Amping out

When creating a woodwind instrument, length and width dictate what type of sounds it can make. This can be seen in the many different styles and designs of didgeridoos all around the country. The didgeridoo was traditionally made by First Australians using termites to hollow out a suitable branch, usually from a eucalypt. Different branches create different sounds. Increasing the width of the instrument increases the amplitude of the sound waves going through it, making it louder. Changing the length also alters the natural wavelengths of sound, producing higher or lower pitches, just like when you put your fingers over the holes on a recorder or a flute.

Diffraction

ULYSSES BUTTERFLY

We can make observations to recognise how light behaves just like sound or water waves when passing through an opening or when two beams of light come into contact. Imagine you're shining a torch into a dark room through a small gap. Although a lot of the light will travel straight through to the other side, some will filter out into the rest of the room – a process known as diffraction. Sound does the same thing, and so does water.

Shiny scales

Diffraction, often along with interference, gives some animals an iridescent sheen. Microscopic nanostructures, like the tiny scales on the abdomens of peacock spiders or on the wings of Ulysses butterflies, form a film that acts as a 'diffraction grating', bending some wavelengths of light that strike it at an angle and making the colours gleam depending on your viewpoint. Interference, refraction and pigmentation can also cause iridesence.

Bend it like a beam

Diffraction occurs when light bends around an object or is forced through a gap. You can see this if you shine a torch through a keyhole or when waves enter a small gap in a wall at the beach, for example. The amount of diffraction depends on the wavelength and the size of the gap or obstacle. If the gap is too big or the waves too small, diffraction will not occur as easily.

DIFFRACTION

Starry spikes

If you look at a star through a telescope, you'll notice it sometimes looks like it has a 'spike' on it – this is called a diffraction spike, and it is an effect caused by the structure of a telescope as it bends the light you see coming from the star. The secondary mirror on a telescope requires internal supports, which interact with the light, causing these diffraction patterns.

SUNSET OVER CENTRAL AUSTRALIA

Diffraction and light scattering are the reasons we see twilight. Red light has the longest wave-length, followed by orange, so these colours bend the most when the sun dips over the horizon, creating vivid sunsets and sunrises.

DID YOU KNOW?

A ring around the moon, or 'moon halo', is often said to mean rain is coming. It occurs when light is refracted from ice crystals in high-altitude cirrus or cirrostratus clouds. As light passes through the crystals, the light is bent by 22 °, forming the halo.

FACT

Both diffraction and refraction bend light. Diffraction is when light bends around an object or through a hole, and refraction is when light passes through mediums like gases or glass.

A story of sunsets

The Tiwi people of northern Australia describe how the sun-woman, Wuriupranili, and the moon-man, Japara, travel at different times across the sky. Each carries a torch of flaming bark, but when they reach the western horizon, they extinguish the flames and use the smouldering ends to light their way as they return eastwards through the darkness of the underground world. Each morning, the fire lit by the sun-woman to prepare her torch of bark provides the first light of dawn. The clouds of sunrise are reddened by the dust from the powdered ochre she uses to decorate her body. Then, the soft, melodious call of Tukumbini, the yellow-faced honeyeater, wakens the people to the duties of another day. At sunset, Wuriupranili reaches the western horizon. But before she returns by underground passage to her camp in the east, she again decorates herself with red ochre, causing the brilliant colours of sunset.

YELLOW-FACED HONEYEATER

The Tiwi people call this species Tukumbini.

Interference

When one wave comes into contact with another, it is called interference. When beach waves intersect at an angle, they simply pass through each other and keep on rolling. But when they make contact, you might notice that the height (amplitude) of each wave is combined, making a bigger wave. This process is called constructive interference. Now you know why when you ride a wave on your boogie board and it catches the wave in front, you get lifted up when the two waves combine. Light waves act the same way.

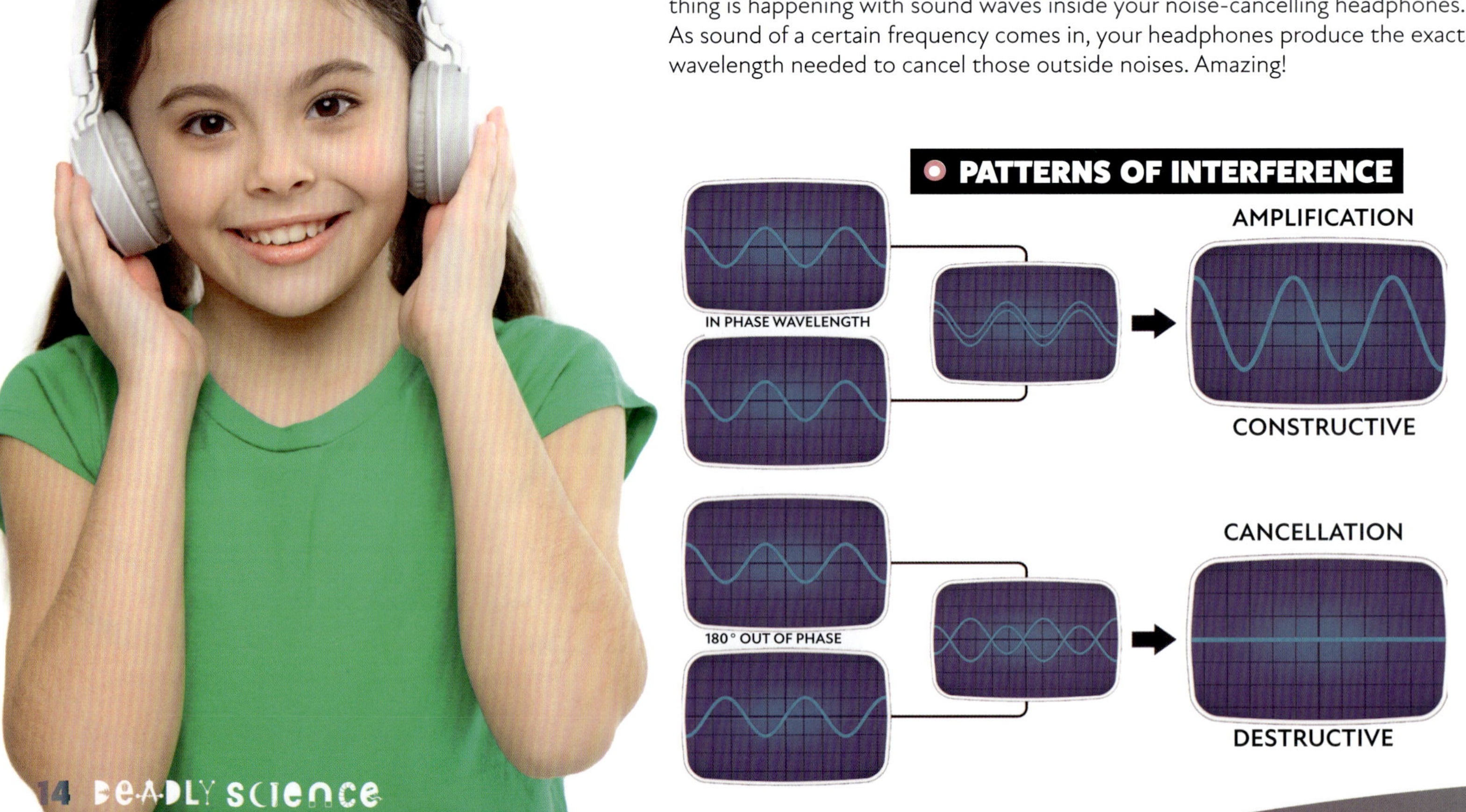

NOISE-CANCELLING HEADPHONES

Cancelled!

Like beach waves, light waves have both crests (high points) and troughs (low points). Sometimes, the crest of one wave collides with the trough of another. They briefly combine, but because they are going in different directions, they cancel each other out, which is known as destructive interference. The same thing is happening with sound waves inside your noise-cancelling headphones. As sound of a certain frequency comes in, your headphones produce the exact wavelength needed to cancel those outside noises. Amazing!

PATTERNS OF INTERFERENCE

YOUNG'S DOUBLE-SLIT EXPERIMENT

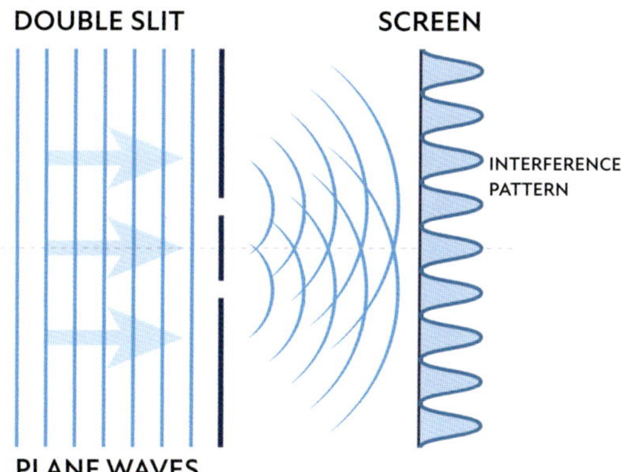

DID YOU KNOW?

It might not seem like it, but when we observe light around us, we are seeing lots of different colours, meaning lots of wavelengths of light. Our eyes don't notice much constructive or destructive interference because there's so much going on that our brains even it out into an average, so we don't see the patterns.

Proving light is a wave

When a scientist makes a claim, they must be able to support it with evidence. In the late 1600s, it would have been very difficult for Christiaan Huygens to convince people that light behaves just like sound or water. Luckily, in 1801, English physicist Thomas Young devised an experiment that helped demonstrate light's wave-like properties. Today, the best piece of evidence we have for the wave model of light still comes from Young's double-slit experiment.

Young's simple experiment is done using monochromatic light – a single colour with a single wavelength so no other light can affect the experiment. The experiment works by shining light onto a surface with two small gaps. Using a single light source, such as a laser, ensures the waves come in a regular pattern without other sources influencing the results. In the diagram at top left, the crest of each wave is shown by a line. Like ocean waves coming to shore, light waves come in as a wavefront. As the waves hit the first screen, light passes through two narrow slits, spreading out (diffracting) on the other side. Imagine you're pouring water down a slope that has only two small openings – this is how the water would flow out onto the other side. As the new waves (one from each gap) interact with each other, two crests meet where the lines cross and a pattern of constructive and destructive interference forms in places where they combine and other places where they cancel each other out.

It's easy to observe these patterns on an optical screen (below left). There, you can see points where the crests of each wave meet (combining their amplitude and making bright spots), and where a crest and trough meet (cancelling each other out, making dark spots). Look at the second part of the diagram and imagine the bright bands represent the crest of each wave. Can you identify where two crests come together? This result can only be described using a wave explanation for light, because only waves combine to cancel each other out.

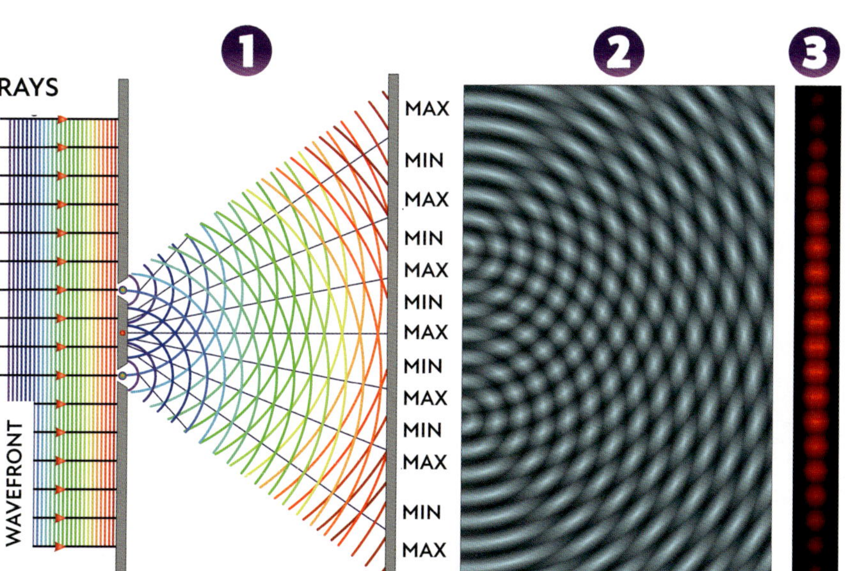

Three views of the same experiment. 1. Shows what happens to the light waves. 2. Shows the effect in water waves. 3. Shows what you'd see on an optical screen with a red-light source.

The spectrum

Much of our understanding of the world comes from our five senses: touch, taste, smell, hearing and sight. But as science has developed, we've discovered new things with the aid of devices that allow us to enhance our own senses, such as using microscopes to examine tiny objects, telescopes to see things far away, microwaves to cook food, thermometers to measure temperature, X-rays to photograph our bones, and magnetic fields to take images of what's inside our bodies! All of these focus on using different wavelengths of light from the electromagnetic spectrum.

COLOUR WHEEL

White light

With all these colours flying around the electromagnetic spectrum all the time, why doesn't our world constantly look like a rainbow? It is partly due to how our eyes work but also because all the colours of light come together to form white light. Think of each colour as having a different energy. When their energies combine, they make a colour full of energy – white light.

Reflection

When light bounces off a surface, some of its energy gets absorbed. Because different colours have different energy levels, when light hits a green object, the object absorbs all the colours except green. Often, the light that is not absorbed bounces off – a process we call reflection. Just as a ball bounces back if you throw it at a wall, when light reflects off a smooth surface, it bounces off at the same angle as it came in. Depending on the surface, this can result in different outcomes. A perfect reflection, such as bouncing off a smooth surface, is called specular reflection. When a rough surface sends light off in lots of different directions, it is called diffuse reflection.

FACT

The longest wavelength belongs to radio waves, followed by infrared, visible light, ultraviolet light, X-rays and, the shortest, gamma rays.

THE ELECTROMAGNETIC SPECTRUM

AM
FM TV
RADAR
TV REMOTE
LIGHT BULB
SUN
X-RAY MACHINE
RADIOACTIVE ELEMENTS

Radio waves
Infrared
Ultraviolet
X-rays
Gamma rays

100m
1m
1cm
0.01cm
1000nm
10nm
0.01nm
0.0001nm

Building Size

VISIBLE SPECTRUM

Atom Size

TRY IT YOURSELF

Shine a red light onto a green surface. As the green surface absorbs red light, it will look almost black.

COLOUR ABSORPTION

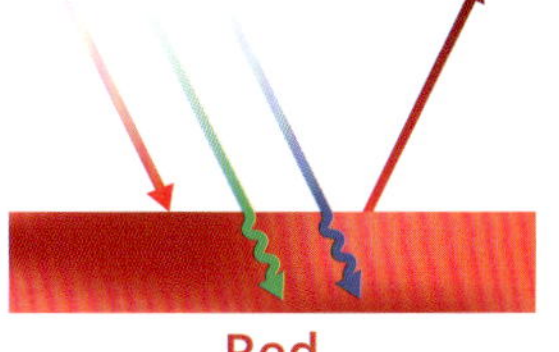

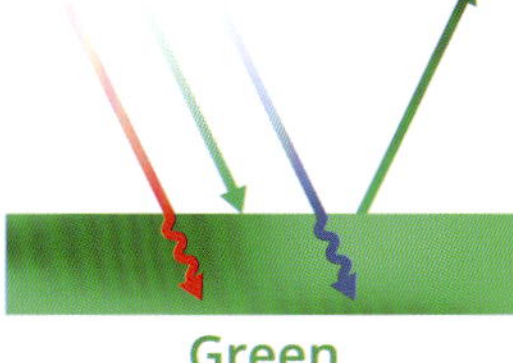

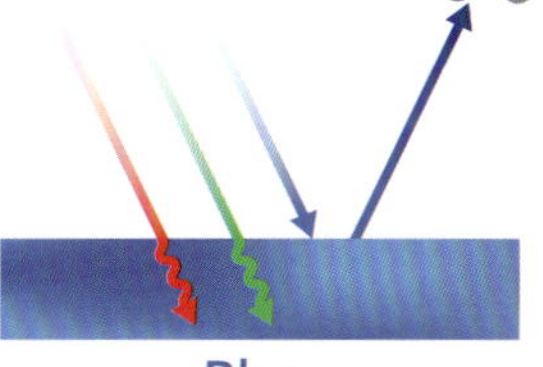

Why we see colour

The image at right is an example of how objects absorb and reflect different colours. It's also why we see things as a certain colour. When you're looking at something red, the object is absorbing every colour except red, letting red light reflect into your eye.

Life in colour

To understand how we see the world, we can start by learning how a TV produces an image. In simple terms, a TV screen is made up of millions of tiny dots known as pixels, each of which is able to create a unique colour. If you imagine a grid of many squares (pixels) you could fill each square with a different colour to produce an image. The number of pixels per inch is called the resolution. The smaller the squares, the higher the number of pixels, and thus the higher the resolution and the clearer the image will be, which is why TV screens with the most pixels cost the most. For example, a 4K TV typically has 2160 rows and 3840 columns of pixels – more than 8 million pixels!

PIXELS

Your brain can tell what colours of light came into your eyes from which direction, and it forms an image based on the many colours different objects have reflected – just like your TV sending different colours from parts of the screen. The next time you're outside, imagine all the colours of light flying around the sky. Then, look at a tree and think about that light hitting a leaf. Leaves use energy from the sun for photosynthesis, but although they absorb most of the light energy, they actually reflect the energy from green light, allowing us to see the green leaves on our trees. We need light to be hitting everything, so it can reflect back, which is why we can see so well in the daylight, when light is bouncing around, but less well in the dark.

DID YOU KNOW?

The sky isn't really blue, you just see it that way. When light enters Earth's atmosphere, it scatters in many directions, like splitting billiard balls. This is called Rayleigh scattering. Blue light has a short wavelength and scatters more, so our eyes perceive the sky as blue.

LOW RESOLUTION (PIXELATED)

HIGH RESOLUTION (CRISP)

Refraction

REFRACTION

Have you ever noticed that when you put your finger or a pencil into water, if you look at it on an angle, it looks like your finger is going in a different direction? This is because light changes direction as it passes from the air into the water and vice versa. We call this process of bending light through a substance like air, glass, or water *refraction*. But how does refraction work?

A HUMAN WAVEFRONT

PRISMS SPLIT LIGHT INTO A RAINBOW

TRY IT YOURSELF
Grab two friends, link hands, and run at different speeds, making sure you don't let go. Eventually, you'll find you have to turn.

Bend it like a beam

When you throw a rock into water, it continues in the same direction, but it also slows down and changes course. Light behaves in a similar way as a wave. If light were a particle, it would just keep going in a straight line. Because waves of light create lots of wavefronts of light coming together in long, connected lines, when part of a wave hits a different medium (like water) and slows, the wavefront causes the wave to turn. Imagine you're holding hands with a few friends and all running toward a sandy beach in one long, wide line, as in the image above. If the group enters the beach at an angle, when the first person hits the sand, that side of the line will slow down. If one side is moving much faster than the other, the only way to keep holding hands would be for some people to turn.

Rainbows and prisms

Because all colours have different wavelengths, colours bend or change direction by different amounts. A prism is a great example. As pure white light comes in (made of lots of wavelengths, like we get from the Sun), it passes through the glass of the prism and bends. Each colour bends different amounts, separating the light into a rainbow of colours. That's exactly how a rainbow gets its colours – when light passes through water at an angle of close to 42 degrees, it separates. That's why a rainbow only appear when there is rain (or water vapour) in the air and sunlight, and also why it forms a perfect arch.

The reason we can never reach a rainbow, of course, is that as we move towards it, the rainbow appears to move back, maintaining the 42-degree angle that makes it visible to us.

DID YOU KNOW?

Pilots in flight can see in more directions, so they sometimes see rainbows that look like a complete circle.

TRY IT YOURSELF

Put a small item in the bottom of the bath and try to grab it out with long tongs. Was it exactly where you thought it would be?

Don't believe your eyes!

Spear fishing is a traditional way of finding food for First Australians. Over generations, they mastered the art, recognising that where they saw the fish and where the fish actually was were different. Because light bends when it enters water, the light reflecting off the fish does not enter the eyes in a straight line. It refracts, a bit like a concave mirror does at the circus. To spear fish, First Australians and other traditional hunters learn to aim slightly ahead of where their eyes tell them the fish is located.

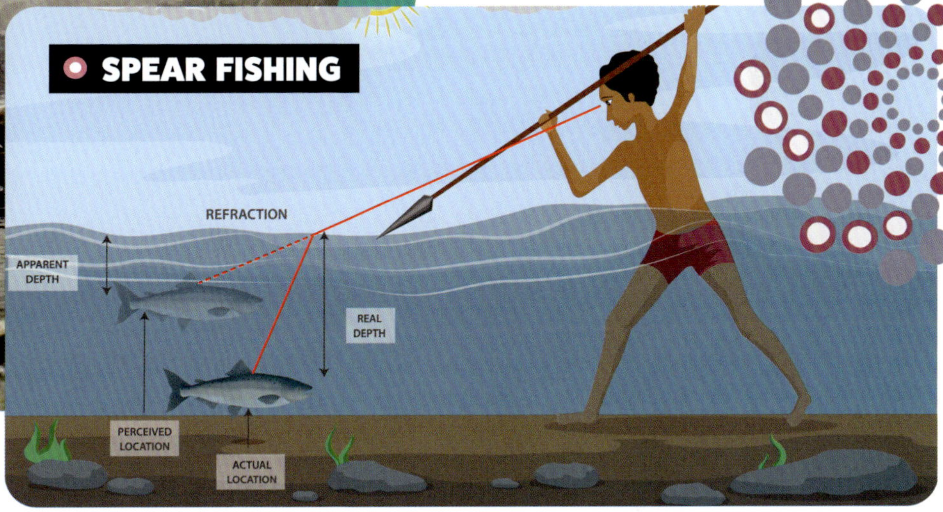

SPEAR FISHING

REFRACTION

APPARENT DEPTH

REAL DEPTH

PERCEIVED LOCATION

ACTUAL LOCATION

SOLAR OVEN

DID YOU KNOW?

Concave mirrors reflect light onto a single point and can be used to make a solar oven. The French Foreign Legion perfected this in the 1870s, allowing soldiers to heat food without smoke revealing their position.

The magic of mirrors

Whenever you look in a mirror, you'll see your reflection. But when you see yourself wearing blue jeans in the mirror, what you're really seeing is light that has hit your jeans, reflecting blue light at the mirror and rebounding the light to your eye. Although you know you're in front of the mirror, your brain is actually seeing someone behind it. Just like a ball bouncing, if the surface isn't flat (as for a curved mirror), the light will bounce off at a different angle. This is why curved mirrors can make you look so funny – they trick your brain into thinking the light has come from somewhere else!

MAGIC MIRROR

Lenses & light

Have you ever wondered how a magnifying glass directs all the light to a single spot? Or how telescopes allow us to see great distances? What about how the zoom on a professional camera works? Or why glasses allow some of us to see more clearly? What all these things have in common is the use of lenses.

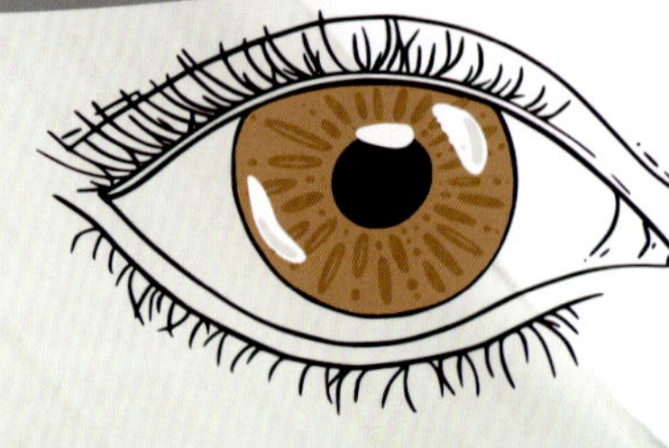

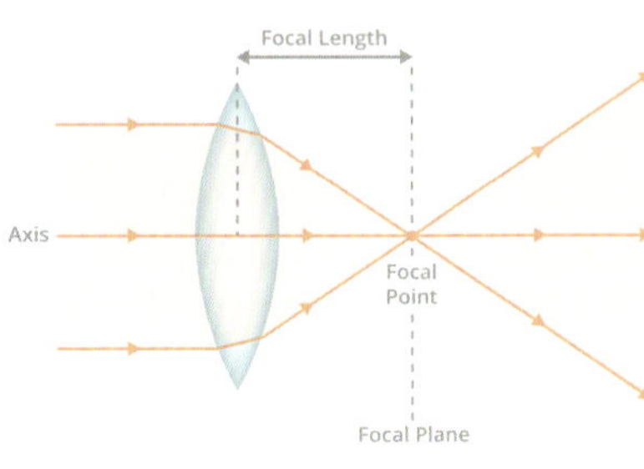

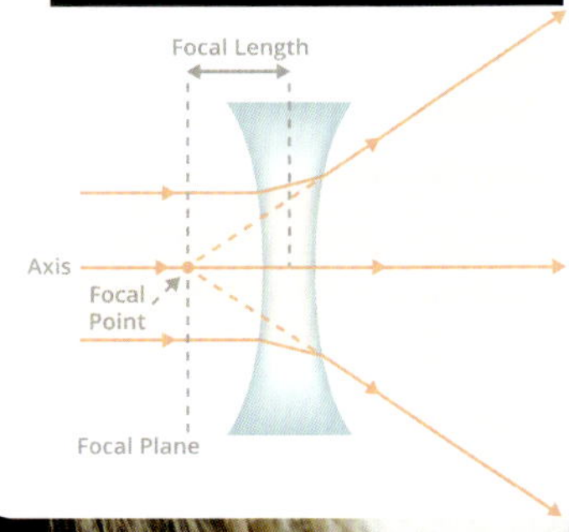

ZOOM LENS

Light benders

A lens is a piece of glass shaped in a way that allows it to bend (refract) and focus light, making big things appear small or small things appear big – just like when we're looking at curved mirrors. Lenses that concentrate all the light into a single point, like a magnifying glass, are known as convex lenses. Lenses that spread the light, like the ones used for car headlights, are concave lenses. Remember how light bends as it passes through glass? The same principle is happening here. Just as a mirror changes the size of something we see, lenses change how our eyes perceive things. The way the light bends as it passes through each lens tricks our eyes into thinking the information is coming from the image, not from the lens.

Using a convex lens can even make images appear back to front or upside down, depending on how far away we are. Take a look at the diagram of the eye (opposite, top right) to see how rays carrying the light from the candle on the left are first reduced in size and directed towards a focal point, just like a magnifying glass, before continuing on upside down!

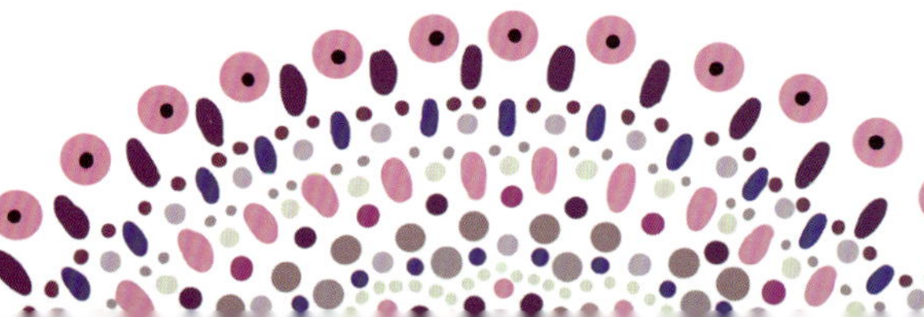

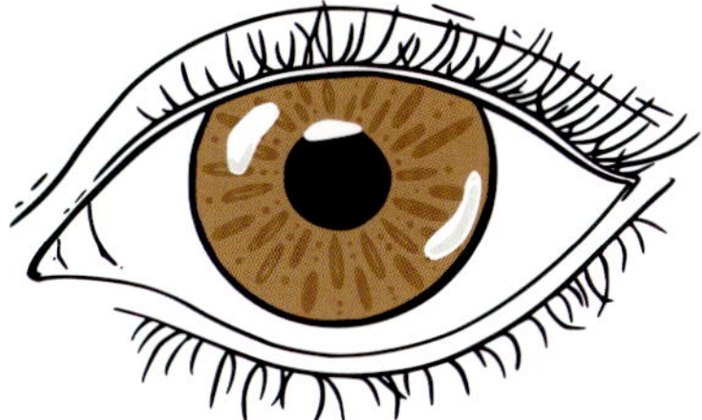

Seeing the light

The human eye is a complex organ that plays a crucial role in how we see and respond to different forms of light. It has many parts, but let's start with the lens. Because the things we often see around us are very big compared to the size of our eyes, we need all that light coming in to be directed to a very small space at the back of our eye, our retina. To do this, the eye needs a convex lens.

You'll notice from the diagram above right that the image of the candle that gets produced at the back of the eye is upside down. But don't worry, our brains are used to this. Your brain will take that data and interpret everything the right way up! Amazing, right?

If the lens in your eye isn't able to focus the light, you'll find it hard to see the candle clearly. Usually, you can fix this by wearing glasses. By putting a different lens in front of the eye, optometrists can adjust how light goes in and make sure everything is clear. This is also why if you don't need glasses but you put on someone else's, everything will look blurry to you.

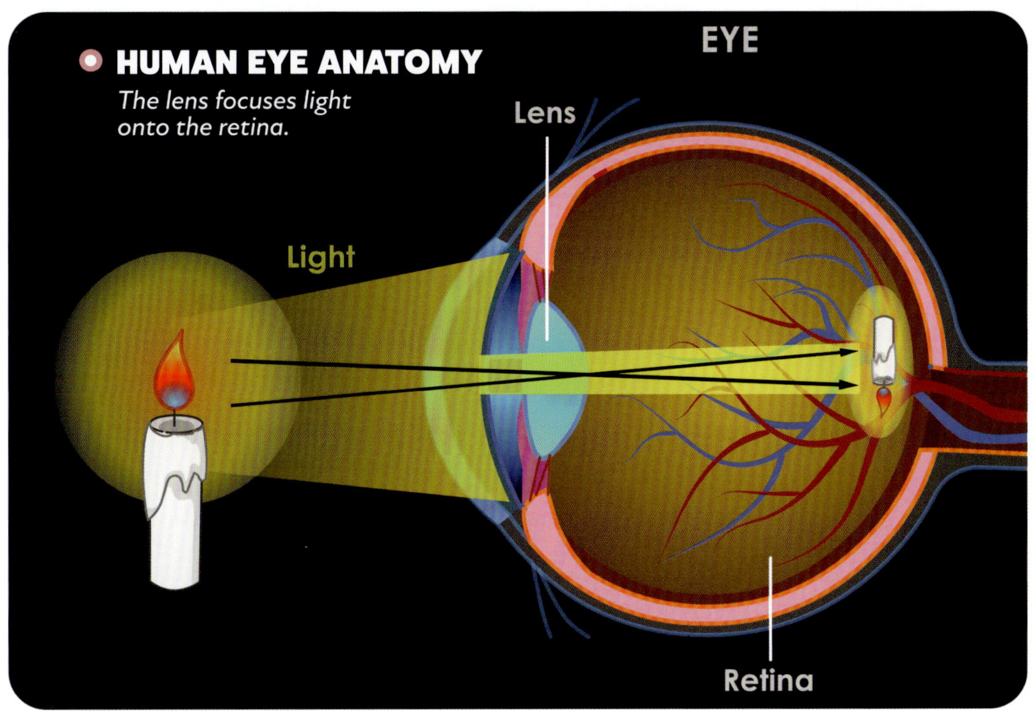

Trick your eyes

1. Take a large clear bowl or a large glass full of water.
2. Draw an arrow on a page and put it on the other side of the glass or bowl.
3. When you look through the glass at the arrow, depending on how far you are from the glass, you should see the arrow pointing in a different direction.
4. Now, move closer or further away from the glass to see how the arrow changes size. At the point where the arrow flips, you won't see anything at all; this is called the focal point.

DID YOU KNOW?

Even if you don't wear glasses, you have a lens in each eye, you see everything upside down, and your eyes have cells that detect only three colours. True!

Life in colour

Although we think we see a spectrum of colours, what we're really seeing is a blend of just three different colours. Each eye has special colour detectors called cones. Cones can only pick up red, blue and green. So how do we see so many other colours?

Colour mixing

As there are so many different tints of blue or red or green, the cones in our eyes can see a range of coloured light, so what matters is how much each cone is detecting.

The graph below shows how each cone receptor allows the eye to work out all the colours of the rainbow. Taking yellow light, for example, the graph shows that it is mainly red, with some green but almost no blue. When yellow light comes into the eye, the eye detects exactly how much of each colour there is, and the brain decides what colour it is seeing. If all of our cone receptors are detecting light simultaneously, the brain interprets that as white light.

TRY IT YOURSELF

Find 3 torches and cover them with red, blue and green transparent cellophane. Shine them on a pale object and see what happens when the colours blend.

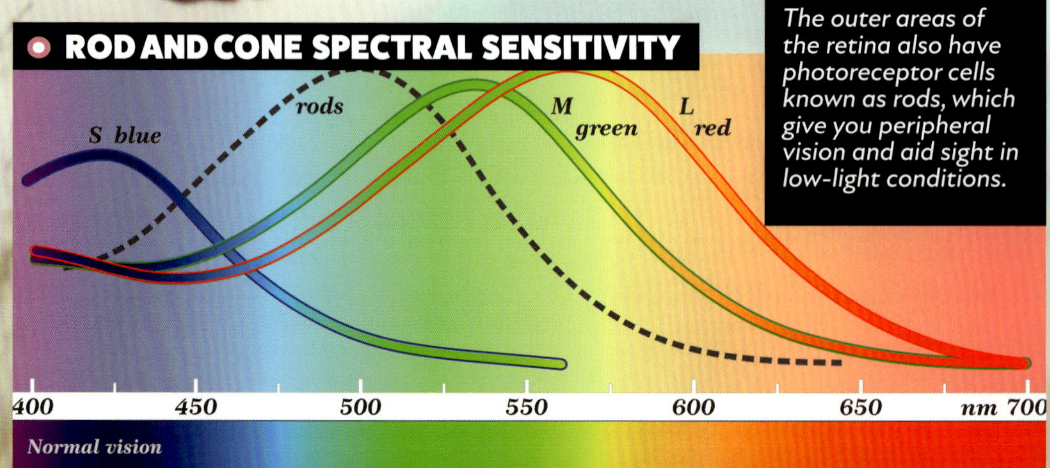

The outer areas of the retina also have photoreceptor cells known as rods, which give you peripheral vision and aid sight in low-light conditions.

ISHIHARA TEST

In this image, everyone should see the number 12.

In the image above, people with normal colour vision will see the number 74, but those who are red–green colour blind will see the number 21.

Colour blindness

Colour blindness is a genetic condition where a person lacks the ability to see the difference between some colours. It can be mild and almost unnoticeable, or, more rarely, extreme enough to mean some people see the world in black and white. For most people, colour blindness barely affects their lives – unless they want to be a pilot or an astronaut.

The main types of colour blindness are red–green and blue–yellow. It can be difficult for colour blind people to distinguish between those colours, so spotting red flowers on a green tree might be a challenge for a person with red–green colour blindness.

The most common test to determine colour blindness is the Ishihara colour test, which uses a series of cards with different spot patterns and numbers on them to figure out what type of colour blindness a person has and how severe it is. The test can be found and completed online.

DID YOU KNOW?

The first TVs and mobile screens made were capable of showing only combinations of red, green and blue. On some old TV screens, if you look closely, you can even make out the pixels.

Beyond RGB

Interestingly, many animals can see colours well beyond just red, green and blue and are able to detect many more variations than humans. Bees and butterflies have four colour receptors, enabling them to see ultraviolet light. A marine invertebrate called the peacock mantis shrimp has an incredible 16 colour receptors and can see ultraviolet, infrared and even polarised light! Humans are good at relying on our other senses and our intelligence to get by, but imagine what the world might look like if we could see in UV or infrared!

AN OWL BUTTERFLY UNDER UV LIGHT

Colour & ceremony

CEREMONIAL PAINT

Traditional Indigenous ceremony involves the use of light and colour, as Aboriginal people decorate their bodies – often to represent moieties, totems of the group, or animals in a story but also to connect with culture by becoming part of the story or dance, rather than just participating in it. Art connects people with Aboriginal culture and is an integral part of the identity of First Australians.

Mixing colours

When we combine colours with light, they become white, so why do paints get darker and form a murky mess when combined? It's because combining light adds more wavelengths together to trick our brains into seeing a different colour, which is called additive colour mixing. Remember that when you see a red object, it's because that object absorbs all the other colours. This is true of paints as well. Red paint absorbs other wavelengths and reflects the red. So, when we combine red paint with a paint that absorbs red, like green, we have subtractive colour mixing – the red paint won't allow all the green to be reflected, and the green paint won't allow all the red to be reflected. The result is something in the middle, like dark green or brown.

The art of storytelling

Indigenous art always tells a story – it's how Aboriginal and Torres Strait Islander culture is kept alive, how lessons are taught, and how children know where to find water or food in different seasons. Paintings, dances and storytelling are used to pass along vital information. That's why it's so important to preserve Indigenous artwork such as cave paintings. They're not just art, they are a living library of First Nations history.

SAND PAINTING

OCHRE

Art on Country

The term Country is often mistaken for just the physical landscape. But when Aboriginal people use the word Country, they aren't just referring to the land but also to their culture and the connection they have with all things on, in, and around their ancestral homes.

Art has always served as a means of connecting with Country through expression, lessons, storytelling, music and dance – all forms of communication. Artworks are made using materials from the land, telling stories about Country while allowing the artist to form a deep connection with what they're trying to say. The story artwork tells isn't just of an image, it is a representation of history and culture.

Artworks can be identified as coming from different areas of Australia not just by what's visually there, but also by the style of the art, the material it is painted on, or the colours used. For example, dot paintings were not common across most of Victoria, where artists relied more on lines and shading. Colours are important because they don't just represent the landscape around the artist, they are also made from it. A number of different methods are used to create paints, depending on where you live in Australia. Usually, you'd first need to find ochre, a natural pigment that's a bit like dried clay. Ochre comes in white, black, reds, yellows, browns, oranges, pinks and purples, and sometimes even in turquoise!

DID YOU KNOW?

Grinding ochre into a powder and then adding animal fats and water to bind it creates a substance not unlike oil-based paint. The process is a bit like adding egg or milk to flour to get a runny, paint-like consistency.

Light cycles

Light affects our lives in many ways. The daily cycle our bodies keep to – driven by the changing from night to day and back again – is called a circadian rhythm. This internal body clock helps our bodies regulate hormones and processes throughout the day. Have you ever noticed that even without a clock or a watch, you can guess the time of day based on when you feel hungry or tired? This is the work of circadian rhythms.

MELATONIN AFFECTS SLEEP

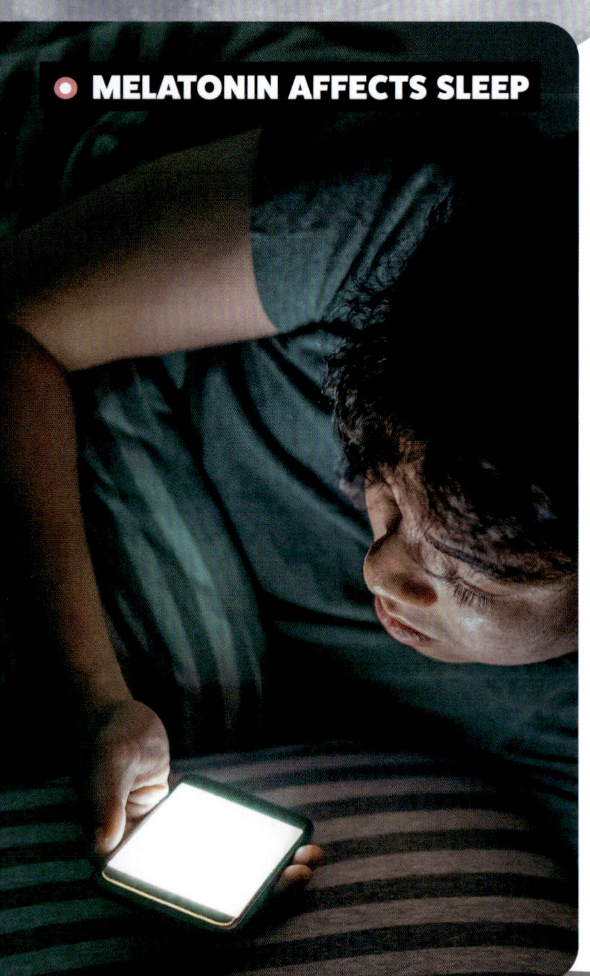

Circadian rhythms

Biological rhythms change throughout our lives. In teenagers, for example, the circadian rhythm is delayed, which is why teenagers love to sleep in! Disrupting your circadian rhythm can make you feel jetlagged and even affect your health, your concentration and your performance at work or school.

Chemical cues

How much light you get also alters your body chemistry and how well your body performs. Skin absorbs vitamin D from sunlight, and your body take cues from day and night cycles. One of the chemicals the human body produces in response to light is a hormone called melatonin. When your eyes detect light, a gland in the brain stops or slows down the production of melatonin, which helps keep you awake. When your eyes don't receive any light, melatonin is produced faster, so you become tired. If you give your eyes a lot of artificial light after dark – by watching screens in bed, for example – it can trick your brain into thinking you're not tired even when your body needs rest.

Melanin and colour

Another hormone, called melanin, affects the colour of your skin, eyes and hair. The more melanin your body produces, the darker your skin will be. Because sun exposure increases melanin (that's what creates a tan), the hotter the sun or the closer to the equator your ancestors lived, the more melanin your skin typically makes as a way to protect against sun exposure.

MELANIN DETERMINES SKIN COLOUR

Star cycles

Over many thousands of years, First Australians observed and learnt from life, land, water and star cycles. By understanding weather patterns, where different animals might be at certain times, and the flowering of edible plants across the seasons, Indigenous Australians are part of their ecosystem, passing information on from one generation to the next through stories, songs and dance. First Australians rely on knowing when each season begins, and to do that, they watch the stars. They search for the appearance of a specific star or constellation in a certain position in the sky. A well-known example is the emu in the sky (top right). By watching the emu make its journey over the year and connecting it with cultural stories, Indigenous people in parts of Australia are able to predict the seasons and when is the best time to hunt for emu eggs.

DARK EMU LYING DOWN

In June/July, the emu's head faces down in the Southern Hemisphere, which is when the male emu sits on the eggs.

Light pollution

Before electricity was invented, the night skies were dark, making it much easier to view constellations and celestial objects. But today, cities stay illuminated all night long and their light pollutes the skies, obscuring some stars. As a result, the International Dark Sky Association (ISDA) has set aside some parks and reserves as designated 'Dark Sky Places' where light pollution must be avoided. In early 2024, Australia had four dark sky parks: Warrumbungle National Park (NSW), River Murray Dark Sky Reserve (SA), The Jump-Up Dark Sky Sanctuary near Winton (Qld), and Arkaroola International Dark Sky Sanctuary (SA).

GREEN TURTLE

Waru is the name for this turtle in the Kala Lagaw Ya language of the Torres Strait Islands.

DID YOU KNOW?

Too much artificial light can harm native animals that rely on moonlight or sunlight to reproduce or navigate. Endangered sea turtle hatchlings find their way to the sea by following the gleam of moonlight on the water. Bright lights near beaches can disorientate them, reducing turtle populations.

Light speed

● ALBERT EINSTEIN

When you turn on the lights at home, do you think about how long it takes for the electric current to travel to the bulb? What about for the light to turn on and reflect off every surface to get all the way back to your eye? Or do you think it is just there when you flick the switch? For a long time, people believed that light was either there or it wasn't. It was not thought of as travelling from one place to another. Discovering that light has speed was a big step in understanding how it works, but just how fast is light?

Light years

Light travels at an incredibly fast 299,792,458 metres every second! That's close to 300,000 kilometres per second – fast enough to circumnavigate the Earth's equator more than seven times in a single second!

Because of how huge the universe is, we can't talk about distances across it like we do car trips. Instead, we talk in terms of light years. A light year is the distance light can travel in one full year. If it can lap the entire planet more than seven times a second, imagine how far light can go in a year. Try keeping that in your head as you think about the closest star to Earth (aside from our Sun), Proxima Centauri, which is 4.25 light years distant – that's 40,208,104,508,468 kilometres away!

Staring at the past

When we look at the stars, we're looking into the past. Some of the stars we're seeing might not even be there anymore. How come? Because they're so far away that it takes a very long time for their light to reach us. When we see Proxima Centauri in the night sky, we're really seeing light that was made by the star and sent our way 4.25 years ago. It has just taken that long for it to get here. Some of our more distant stars are much further away. One of the most distant stars we can see without a telescope is in the constellation Cassiopeia, more than 16,000 light years away. If you imagine you were using lasers to signal to someone looking out at Earth from Cassiopeia, they would receive the light you sent them in 16,000 years!

Speeding into the future

The Theory of Special Relativity ($E = mc^2$) is one of the most famous scientific equations about light, and it came from no other than German theoretical physicist Albert Einstein in 1905. It states that E (the total energy of something) = m (its mass) × c^2 (the speed of light x the speed of light). This equation tells us something very important: nothing can travel faster than the speed of light – not even light itself. Instead, as something gets faster, all of its extra energy goes into its mass. Einstein's Theory of Special Relativity helps explain how matter formed in the universe and inspired entire new fields of study within physics, paving the way for modern electronics, nuclear power, and GPS navigation.

DID YOU KNOW?

Light behaves differently depending on how we measure it, which is why some scientists call light a 'wav-icle', because it seems to behave like a wave and a particle. This is called the Wave–Particle Duality theory. Some people believe a mysterious phenomenon known as ball lightning is an example of this duality, and physicist Nikola Tesla once claimed to have created it in his lab!

BALL LIGHTNING

Particles of light

An experiment known as the Photoelectric Effect proves that light sometimes behaves as a particle. As light has a level of energy directly connected to its colour, if we shine light on a special metal surface, it can give the electrons on that surface enough energy to jump across a small gap and create an electrical current. This action is only supported by a particle model of understanding because of the way waves provide energy.

PHOTOELECTRIC EFFECT

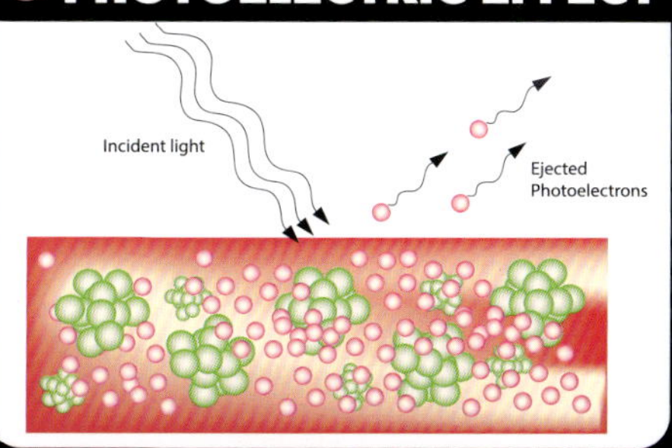

For example, when you're heating a pot of water, as long as the hotplate is more than 100 °C, the water will eventually boil because heat is absorbed bit by bit until it reaches boiling point. This is also how waves provide energy. According to the wave model of light, if we shone light with very little energy on a special surface and waited, there should eventually be enough energy for the electrons to jump across. But that isn't what we find. Instead, we see that electrons either do or do not jump across the gap, supporting the idea that tiny balls of light energy, called photons, have collided with the electrons, either giving them enough energy to make the jump or not – and supporting a particle theory of light.

Where to from here?

Although scientists can explain how light works, predict its behaviour and create entire industries and products based on its use, such as LEDs and solar panels, we still don't know exactly what light is, only how it behaves. For some, that might be a little disappointing, but it should be exciting.

One of the amazing things about science is how our understanding of the world around us grows as we learn new things. Light and colour are examples of scientific concepts that a clever young mind could dive into and discover something no one else has thought of before. Who knows, maybe you'll be the next du Châtelet, Huygens, Edison, Tesla, Newton or Einstein!

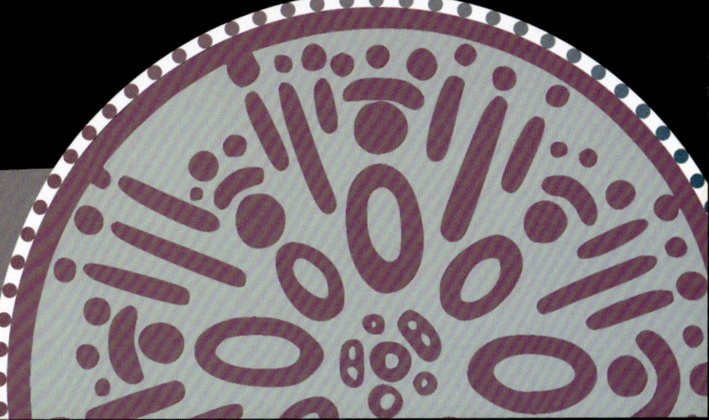

Chromatography: Kalkarindji School on Gurindji Country, NT

Students at Kalkarindji School on Gurindji Country in the Northern Territory performed a chromatography activity to learn about pigments and the components of colours.

Materials

- Paper towel cut into long strips about 3–5 cm wide
- Felt-tip colouring pens in assorted colours and brands
- A cup or mug half-filled with water

Observations and predictions

Observe how colours react to the water being soaked up by paper towel. How do you think certain colours will react? Will all of the colours react the same way?

GETTING STARTED

MAKING OBSERVATIONS

MAKING COMPARISONS

Method

STEP 1
Using one of the pens in a colour of your choice, draw a dot (about the size of a big full stop) a little bit up from the bottom of one of the strips of paper towel.

STEP 2
Place the very bottom of the strip of paper towel in the water, being careful to not get the dot itself wet. Leave it in for a few seconds until the water seeps in and the ink starts to split into colours. Then take it out and leave it flat on a piece of blank paper.

STEP 3
Repeat steps one and two, using different colours on each new strip of paper towel.

WARM COLOURS

COOL COLOURS

STEP 4
Observe what happens to the colours over time.

Questions

- Did the colours all react the same?
- If not, why do you think the colours split differently?
- If you used two pens of the same colour but from two different brands, were the split colours still the same?
- How do you think the water moves up the paper towel?

What this shows

This shows that the colours we see are made up of primary colours added together, a process known as additive colour mixing.

We can also see that most colours easily separate into other colours, or pigments, with the lighter colours moving further up the paper towel. This means that lighter colours are, in fact, lighter in weight, too. Scientists call this chromatography, and it is used to learn more about colours and what things are made of.

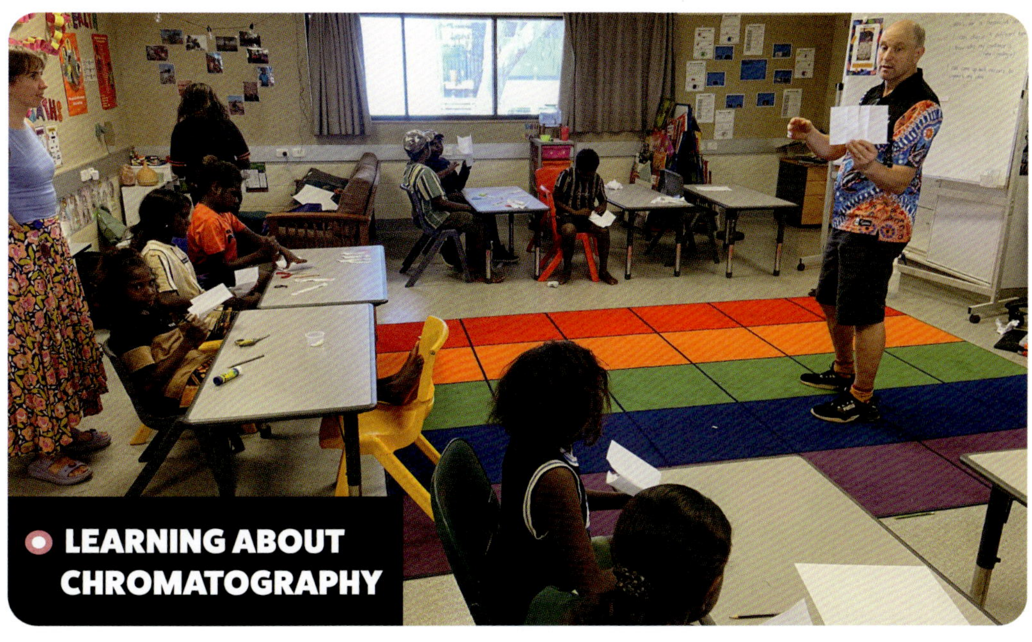
LEARNING ABOUT CHROMATOGRAPHY

First published in 2024

52–54 Turner Street
Redfern NSW 2016

editorial@ausgeo.com.au
australiangeographic.com.au

Series Editor: Corey Tutt
Text: Paul Briggs
Illustrations: Karina Jeffrey–Garawa Aboriginal Designs; Mim Cole, Mimmin Design

Commissioning Editor: Karin Cox
Layout: Karin Cox
Proofreader: Serene Conneeley
Cover Designer: Paul Hodge
Print Production: Andy Franks

AUSTRALIAN GEOGRAPHIC
Managing Director: David Haslingden
Editor: Karen McGhee
Head of Digital Content: Charlie Page
Licensing and Publishing Manager: Tom Bates

Printed in China by C&C Offset Printing Co. Ltd.

A catalogue record for this book is available from the National Library of Australia

Picture credits

Front Cover: Twinsterphoto/Shutterstock (SS); Mim Cole, Mimmin Designs; Karina Jeffrey–Garawa Aboriginal Designs; Kucharski K. Kucharska/SS; Liga Alksne/SS; Vojce/S; M-Three/SS; Reshetnikov_art/SS; Wakcomca/Dreamstime (DT). **1:** DeadlyScience. **2:** Tulip D/SS; ChameleonsEye/SS; CFH DESIGN/SS. **3:** Nicole Patience/SS; Dimitrios Karamitros/SS; Andrea Dante/SS. **4:** Ryan Hoi/SS; Elena11/SS. **5:** Petar B photography/SS; D. Kucharski K. Kucharska/SS; RugliG/SS; Vojce/SS. **6:** Oka2011/Dreamstime (DT); WMC/Public Domain. **7:** Klemzy/SS; Harbucks/SS; NDAB Creativity/SS; glenda/SS; Jose Luis Stephens/SS; Karina Jeffrey–Garawa Aboriginal Designs. **8:** M-Three/SS; ChristinaRichards/DT. **9:** Pressmaster/DT; Iana Surman/SS; Mifid/SS; BNP Design Studio/SS. **10:** Nicku/DT; Karina Jeffrey–Garawa Aboriginal Designs; Duda Vasilii/SS; Dream01/SS. **11:** Wakcomca/DT; Rdalland/DT; Mim Cole, Mimmin Designs. **12:** Wako Megumi/SS; Folgen/Pixabay Roberto Lo Savio/SS; kwest/SS; Karina Jeffrey–Garawa Aboriginal Designs. **13:** Gunnar Grindvol/SS; Karina Jeffrey–Garawa Aboriginal Designs; AlecTrusler2015/SS.**14:** Alex Pix/SS; Fouad A. Saad/SS; Just Dance/SS. **15:** Dream01/SS; Delpixel/SS; Fouad A. Saad/SS. **16:** Liga Alksne/SS; Vector Mine/DT. **17:** petrroudny43/SS; Kaspars Grinvalds/SS; MDV Edwards/SS; Karina Jeffrey–Garawa Aboriginal Designs; Adpragus/SS. **18:** Pat Hastings/SS; Reshetnikov_art/SS; Tint Media/SS. **19:** EdArtworks/SS; Prachaya Roekdeethaweesab/SS; Nandalal Sarkar/SS; Jayakri/SS; agsaz/SS. **20:** Zurijeta/SS; Karina Jeffrey–Garawa Aboriginal Designs; petrroudny43/SS; Mim Cole, Mimmin Designs; Anna Nikonorova/SS. **21:** grayjay/SS; Karina Jeffrey–Garawa Aboriginal Designs; NAB/SS. **22:** chuanpis/SS; art-foto/SS; Oleksii Liebiediev/SS; Fouad A. Saad/SS. **23:** chomplearn/SS; WMC; Wellcome/WMC; Purfect_photo/SS; Oleksiy Mark/SS. **24:** ChameleonsEye/SS; Karina Jeffrey–Garawa Aboriginal Designs; LN team/SS. **25:** Derind99/DT; Karina Jeffrey–Garawa Aboriginal Designs; rweisswald/SS. **26:** Netscape/SS; Viktoria Kurpas/SS Independence_Project/SS; ivansnap/SS. **27:** Taylor Wilson Smith/SS; Amirreza Kamkar/IAU OAE/WMC; Karina Jeffrey–Garawa Aboriginal Designs. **28:** Elena11/SS; Jyotiai/SS Aree_S/SS; Formatoriginal/SS; Karina Jeffrey–Garawa Aboriginal Designs. **29:** Amanita Silvicora/SS; Karina Jeffrey–Garawa Aboriginal Designs; KKT Madhusanka/AdobeStock. **30:** romirijpg/SS; DeadlyScience. **31:** DeadlyScience; myboys.mejp/SS.
Back cover: Mim Cole, Mimmin Design.